Contents

Page

LIST OF ILLUSTRATIONS..iii

LIST OF TABLES.. iv

ABSTRACT.. v

INDIRECT THREATS TO THE HUMAN WEAPONS SYSTEM...1
    The Human Weapons System...1
    BIOLOGICAL HAZARDS IN THEATER 2
    CHEMICAL HAZARDS IN THEATER 2
    PHYSICAL/THERMAL HAZARDS IN THEATER 3
        Noise..3
        Light...4
        Non-ionizing and Ionizing Radiation 4
        Heat Stress Injuries..5
        Ergonomic Hazards 6
        Defending the HWS..7

DEFENDING THE HWS USING THE BIOENVIRONMENTAL ENGINEER
    CAPABILITIES AND ORM...8
    Occupational Health Controls and Surveillance...8
    Integration of BEE in Operational Risk Management...9
    ORM on Biological Hazards 11
    ORM on Chemical Hazards..12
    ORM on Physical/Thermal Hazards...12
        Noise..13
        Light...13
        Non-ionizing and Ionizing Radiation 14
        Heat Stress Injuries..14
        Ergonomic Hazards 15
        Minimizing the Risks to the HWS..15

HISTORY OF ILLNESS IN COMBAT..17

General Effects of Indirect Attacks on U.S Forces in History 17
Gulf War Hazards 18
    Biological Illness 18
    Chemical Illness 19
    Physical/Thermal Illness 20
Recent Employment of Preventive Health Measures 21

FOCUSING PREVENTIVE MEDICINE EFFORTS 23
Coordinated Efforts 23
Failures in Focusing PM Efforts 23
    Biological – Improper sanitation and hygiene 24
    Chemical – Inhalation hazard exposures 24
    Physical/Thermal – DU exposures 25
Successes in Focusing PM Efforts 25
    Biological – Pest management 25
    Physical – Heat stress 26
    Balkans – PM Focus at its best 26
    Need for Doctrine Regarding Preventive Medicine and the BEE 27
    Need for a supported PM Program 28

RECOMMENDATIONS 29
Advocating an active PM program 29
    Defense to Biological Threats 29
    Defense to Chemical Threats 30
    Defense to Physical/Thermal Threats 30
Command Support 31

CONCLUSIONS 32
Preventive Defense of the HWS 32

GLOSSARY 37

BIBLIOGRAPHY 39

List of Illustrations

Page

Figure 1 1

Figure 2 10

List of Tables

Page

Table 1 6

Table 2 34

Abstract

Historically, for every one soldier who was a battle casualty 10 other soldiers were unable to fight due to the indirect attacks. These indirect attacks came from biological, chemical, and physical/thermal fronts. Using the Gulf War as a case study, this paper investigates indirect attacks on the Air Force's most valuable resource, people, also known as the Human Weapons System (HWS). This system is defined as the combined physiological make up of individual airmen. Threats to the HWS are identified and the author contends that the HWS like any other system in the Air Force must be defended and maintained to optimize its effects in the combat theater. The author explains that to do this; commanders must integrate the Bioenvironmental Engineer (BEE) into the Operational Risk Management (ORM) process "…to maximize operational capabilities while minimizing risks…"[1]. This integration is necessary because when the BEE is included as part of the ORM in beddown planning, on advance teams, and in the actual operations, commanders can obtain useful information on minimizing the medical risks to their forces, the HWS. For instance, the BEE can provide surveys to identify and defend against biological, chemical, and physical/thermal attacks. Indirect attacks encountered in the Gulf War are introduced and an analysis is done to determine successes and failures of defending the HWS against indirect attacks in the Gulf War. Lastly, recommendations and conclusions are made regarding the BEE's current role and capabilities, and his future usefulness in combat.

Notes

[1] AFI 91-213

Part 1

INDIRECT THREATS TO THE HUMAN WEAPONS SYSTEM

The Human Weapons System

Historically, for every one soldier who was a direct battle casualty 10 other soldiers were unable to fight due to indirect attacks on their health.[1] The health of soldiers refers to their psychological, physiological, sight, hearing, respiratory, digestive, and gross and fine motor systems (see figure 1). All of these things combined can be viewed as the Human Weapons System (HWS).

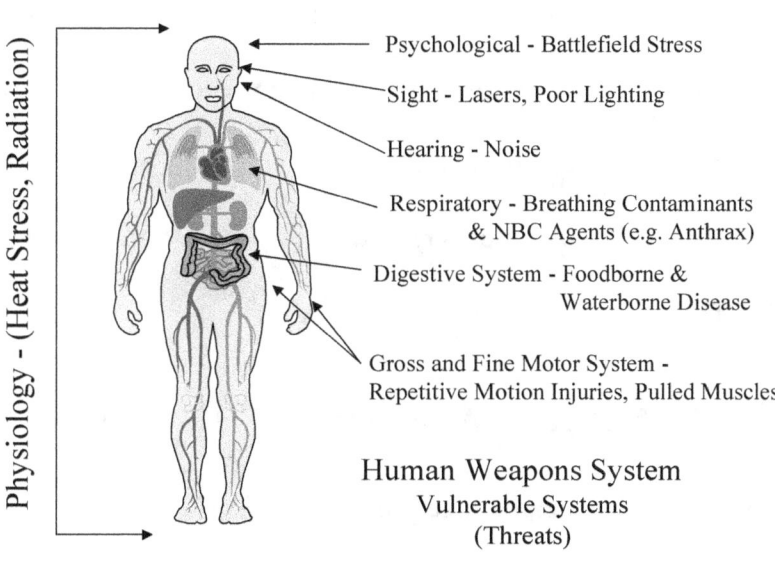

Figure 1

The commander is rightly concerned with the direct threats to the HWS such as gunshots, shrapnel, and explosive forces. Unfortunately, it is easy to get caught up in the obvious threat of enemy action and fail to consider indirect threats. This is because indirect threats to the HWS take more subtle forms. These indirect threats come on biological, chemical, and physical/thermal fronts.

BIOLOGICAL HAZARDS IN THEATER

A wide range of naturally occurring (or endemic) disease is found in every region of the world and poses threats to both the digestive and the respiratory systems of the HWS. Some regions pose serious health threats due to lack of hygiene, water treatment facilities, and poor medical infrastructure. Animals and insects in most parts of the world can also transmit microbes or viruses through bites. Expected biological diseases in foreign theaters range from dysentery and cholera that are acute bacterial diseases to malaria and dengue fever, which are diseases transmitted by mosquitoes.[2] These diseases, along with countless others, are ever present in deployed theaters and warrant caution to prevent the vulnerable HWS from succumbing to their symptoms. The military impact of insect borne disease alone accounted for 68 percent of all hospital cases in the Gulf War, 64 percent in Vietnam, 67 percent in Korea, 87 percent in World War II, and 99.4 percent of the 450,000 men struck down by louse borne typhus in Napoleon's march on Russia (1812-1813).[3]

CHEMICAL HAZARDS IN THEATER

Routine non-enemy chemical hazards take the form of fuels, detergents, and exhausts in the combat theater. Exposures to routine chemicals found every day on the job in the operational environment can have minor to severe health effects on the sight, respiratory and physiology (skin) systems. Safeguards must be designed to create barriers between toxic substances and the HWS. Without properly designed systems and periodic inspection of these safeguards, the operational capacity of the air base could be jeopardized. For instance, a typical maintenance task such as painting can result in vapor exposure to an entire hangar of airmen causing them to suffer immediate signs of nausea and headache. The resulting degradation in the work force could easily effect the Flying Mission Capable (FMC) rate and reduce available sorties.

PHYSICAL/THERMAL HAZARDS IN THEATER

Noise, light, non-ionizing and ionizing radiation, and heat stress injuries are among the most likely physical hazards for the HWS in the combat theater. Physical hazards attack specific components of the HWS or the entire physiology of an individual. These perpetual hazards are ever present on the flightline, in the maintenance shops, and even in administrative type areas.

Noise

Nerve deafness and tinnitus (ringing of the ears) is a possible illness resulting from the combat environment. This impacts the hearing system of the HWS. Flightline noise is likely to be more continuous in the event of increased operations tempo. Location of barracks, latrines, and dining halls as well as the flight pattern routes can make a dramatic impact on the intensity and duration of exposure. Effects of noise exposure may be temporary or permanent, however either way even partial deafness can lead to difficulty in communication in this communication intensive environment. Mission degradation at the tactical level of combat includes difficulty in listening to critical radio transmissions and mistakes in interpreting vital information such as UXO coordinates or battle damage assessments (BDA) for the runway, utilities, and buildings.

At the Operational level, hearing impaired individuals may have trouble with hearing verbal directions and warnings while suited in Mission Oriented Protective Posture (MOPP) gear.

Light

Poor lighting conditions lead to a variety of disabling symptoms such as headaches and eyestrain. This impacts the sight system in the HWS. Some demands like drafting plans for beddown and other administrative type duties are extremely difficult in the temporary shelters provided in expeditionary theaters. These workspaces need to be monitored to ensure that lighting is optimized to avoid chronic ocular illness and to ensure the HWS is fully functional and available for other duties when they are needed.

Lasers (light amplification by stimulated emissions of electrons) and other high intensity lighting devices have become common place in the combat theater. They are used for measurements, pointers and targeting. Lasers can pose an ocular (eye) and cutaneous (skin) hazard to the sight system and physiology of the HWS. Unprotected exposure to the high intensity laser light can vary from harmless to serious burns or permanent blinding. Exposures to even laser pointers can render permanent blindness in the effected parts of the eye.[4] Though loosing just one HSW to a laser exposure may not impair an entire operation, the individual may be a one-of-a-kind talent such as an experienced expert on avionics that could have greatly enhanced the performance of an entire flying wing.

Non-ionizing and Ionizing Radiation

Close proximity of communications transmitters and microwave stations increase the likelihood of non-ionizing radiation. Exposure to non-ionizing radiation primarily affects the sight (eyes) and physiology (skin) of the HWS.[5] Non-ionizing radiation has enough energy to vibrate molecules causing cellular and tissue damage.[6] Because this hazard is invisible, only qualified personnel should be allowed within prescribed distances of radiating equipment.

Like the non-ionizing radiation caused by radiating devices, ionizing radiation also poses serious health risks in the combat theater. Ionizing radiation is caused from substances rather than devices. Such things as Depleted Uranium (DU), used in reactive armor and armor piercing munitions and Americium and Radioactive Nickel, used in NBC detection devices are possible sources of ionizing radiation on the battlefield. Radioactive ballast is used in some aircraft and can be a threat to emergency personnel responding to an aircraft accident. Acute, high dose exposure can produce acute radiation syndrome, which can lead to death in one day to two weeks, depending on the intensity of the exposure.[7]

Heat Stress Injuries

Heat stress injuries come from a variety of sources in the combat theater. This impacts the physiology of the HWS. In desert regions, sunburn, heat exhaustion, and heat stroke are serious considerations and can cause the HWS to malfunction. These concerns have been exacerbated by the need to wear MOPP gear in the combat theater.

In the 1998 RAND report, it states that "In the next 10 to 20 years, there is a distinct possibility that the United States will be involved in a conflict in which the adversary will use chemical or biological weapons."[8] The introduction of man-made disease, such as weaponized Anthrax to the battlefield, place additional urgency on defining, quantifying and predicting effects of NBC events in defense of the HWS. The Iraqi forces had a considerable quantity of Anthrax on hand during the Gulf War. Even though the weapon was never used, the threat of its use forced airmen and soldiers to wear protective gear that caused additional thermal (heat) stress on the HWS. Task multipliers must be applied for the wear of the MOPP gear (see Table 1) to compensate for the additional heat stress on the wearer. These multipliers greatly delay sortie generation times.[9]

Work Levels

Work Level	Task Example	Task Time Multiplier
Light	inspection	1.31
Moderate	refueling	1.5
Heavy	munitions loading	4.24

RAND

Table 1

With the threat of chemical and biological munitions, the danger of heat casualties due to MOPP gear is increased while work levels decrease as shown in Table 1. "It does not take long for people doing heavy work to become heat casualties. It takes only 32 minutes of heavy work without cooling to reach a stored heat level of 160 kilocalories (Kcal), a point at which 50 percent of the population will suffer heat stroke."[10]

Ergonomic Hazards

Repetitive Motion Injuries (RMI) are common in nearly every working environment. The possibility of a RMI in the combat theater is just as probable. This impacts the gross and fine motor systems of the HWS. Even lifting boxes to load freight can cause lower back injury serious enough to incapacitate the HWS. The rush and urgency to complete a given task during a contingency may increase the likelihood of a pulled muscle or a strained back. Risk factors for low back injuries that result in disability include heavy repetitive lifting and pushing and pulling, as well a exposure to industrial and vehicular vibrations, all of which can be found in the combat theater.[11] Statistics are difficult to obtain on these type injuries because many are never reported. "It should be noted that a definitive diagnosis cannot be reached in 85 percent of patients with

6

low back pain."[12] However, it is reasonable to consider that airmen operating with pulled muscles are not at their full physical potential.

Defending the HWS

The combat commander can identify and defend against the aforementioned threats by utilizing the skills and equipment provided by the BEE career field. This career field can identify, analyze, and quantify threats to the HWS as well as provide preventive medicine in the form of surveys, selection of personal protective equipment, water testing, and compliance monitoring. The following investigates how the BEE can use occupational health controls and surveillance and industrial hygiene to minimize the risks to the HWS. This industrial hygiene methodology must be integrated in the Operational Risk Management (ORM) process to defend each of the critical subsystems of the HWS.

Notes

[1] Lt. Gen. Bigert, AFMC Leading Edge, November 1999, p. 5.

[2] Benenson, Abram S., Control of Communicable Diseases in Man, American Public Health Association, 15th Edition, 1990, 17-20

[3] Bioenvironmental Engineering Readiness, Vol. 2 NBC Operations, Brooks AFB, Feb 98, p. 6-1.

[4] McCunney, Robert J., A Practical Approach to Occupational and Environmental Health, 2nd Edition, Little, Brown and Company, Boston, MA. 1994, p. 438-9.

[5] Ibid, p. 438.

[6] Ibid, p. 438.

[7] Ibid, p. 438.

[8] Chow, op cit, vii.

[9] Brian G. Chow, "Air Force Operations in a Chemical and Biological Environment", RAND, 1998, 146.

[10] Ibid, 129.

[11] Op cit. McCunney, p.166

[12] Ibid, p.167.

Part 2

DEFENDING THE HWS USING THE BIOENVIRONMENTAL ENGINEER CAPABILITIES AND ORM

It is essential that Joint Chiefs of Staff take aggressive action now to help protect the health and safety of deployed personnel and facilitate possible medical follow-up. The Military Departments have the safety and health expertise to prevent injury and illness. Documentation of hazardous exposures is needed to allow assessments of any health consequences, which may develop following deployment[1].

—Sherri W. Goodman Deputy Under Secretary of Defense (Environmental Security)

Occupational Health Controls and Surveillance

Occupational health controls and surveillance are required to ensure that the HWS is separated from the most hazardous conditions in the workplace. The civilian sector has mandated that employers provide health protection for their workers under the Occupation Safety and Health Act of 1970 (OSHA). The military has followed suit accepting and tailoring their job safety and health plans for non-wartime activities. However, in general, deployed forces often are unnecessarily placed at risk by not incorporating the proper health surveillance plans. For example in the Gulf War, in the mad rush to accomplish painting of equipment many individuals were unnecessarily exposed to paint vapors because of failure to comply with health surveillance inspections.[2]

Since the forth century BC, occupational illnesses were recognized as force reduction agents. In the case of lead toxicity, Hippocrates discovered occupational illness in the mining industry.[3] Other men of antiquity such as Pliny the Elder, a first century AD Roman scholar, detected respiratory hazards to workers that used zinc and sulfur and took perhaps the first engineering initiative to design a face mask (from an animal bladder) to protect individuals from exposure to dusts and fumes. This type of problem solving methodology, called Industrial hygiene, is precisely the charge of the BEE.

The necessity for industrial hygiene carries over into the combat theater. Maintenance, fuels, munitions, avionics, civil engineering and other functions present the threat of repetitive motion injury (RPI), inhalation and dermal (skin) exposure to toxins, thermal (heat stress) injuries, radiation exposures, and noise hazards. The culmination of other stressors, such as lack of sleep and psychological stress, make individuals even more susceptible to physical illnesses.[4] For these reasons, the need to analyze, identify, and measure workplace hazards that cause sickness, impaired health and/or significant discomfort to the HWS continues in the combat theater. The commander can utilized the unique expertise and tools of the BEE career field to perform the industrial hygiene function. Using this industrial hygiene methodology, the BEE can be integrated in the Operational Risk Management (ORM) process to defend each of the critical subsystems of the HWS.

Integration of BEE in Operational Risk Management

Commanders must perform Operational Risk Management (ORM) "… to maximize operational capabilities while minimizing risks…"[5] Without the BEE shop on station both prior to and during operations, the commander is left with an incomplete picture of the risks to his forces. This chapter will demonstrate how the Bioenvironmental Engineer (BEE) career field assists the commander in each step of the ORM process as it applies to the health of his forces or

HWS. In order for a commander to conduct ORM on the HWS, he must:

1. Identify the hazard
2. Assess the risk
3. Analyze risk control measures
4. Make control decisions
5. Implement risk control
6. Supervise and review[6]

The commander need not be alone in the process of identification of hazards to the HWS.

By using the BEE shop, the commander can not only get accurate quantitative and qualitative

information for the first part of the ORM process, but can also obtain valuable options in the

defense of the HWS. As risks to vulnerable subsystems of the HWS are identified, BEE staff can

perform surveys, Respiratory Protection (RP) fit testing, and design controls to identify high-risk

conditions. (See Figure 3)

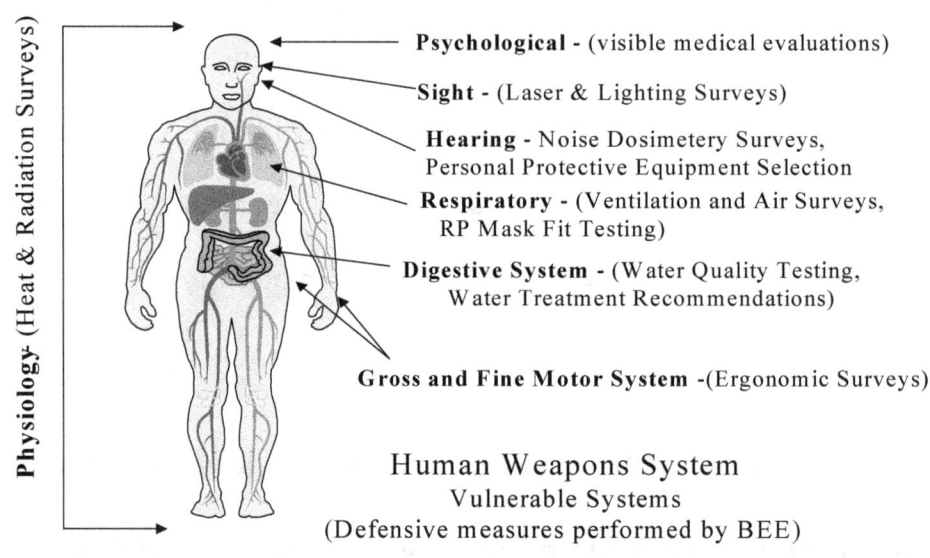

Figure 2

This information can be analyzed (Step 2 of ORM) by the BEE and options for risk control measures can be performed (Step 3 of ORM). Control decisions may then be selected by the commander (Step 4 of ORM) and implemented by the appropriate shop supervisors (Step 5 of ORM). Lastly and often most importantly, the BEE can supervise and review the risk control measures through periodic inspections (Step 6 of ORM). The following enumerates the services the BEE shop can provide to safeguard the HWS against indirect threats.

ORM on Biological Hazards

Drinking water is a major source of biological contaminants that causes illness, such as Cholera and other bacterial disease. They are usually contracted by drinking water contaminated with feces or vomitus or ingestion of foods prepared with dirty water, dirty hands or that has had contaminated flies on it.[7] For these reasons, drinking water (potable water) quality testing and treatment as well as water vulnerability assessments are part of the BEE's repertoire of job skills. This shop has or can procure all of the necessary equipment to ensure that the deployed personnel's drinking water is safe. Additional capabilities include the treatment of sanitary sewage and hygiene as well as determining the viability of water reservoirs such as ponds, lakes and rivers to mitigate contamination. The BEE has equipment to determine flow rates, temperatures and oxygen content of waterways to keep these water sources from stagnating and becoming a health threat.[8] With these capabilities the primary biological risks can be identified, assessed, analyzed, and risk control measures can be taken (the first three steps of ORM). The remaining steps of control decisions, implementation, and supervision may be delegated by the commander to the Civil Engineer Squadron (CE) or the BEE shop.

ORM on Chemical Hazards

Risk management of the routine chemical hazards found in the combat theater environment must be identified, analyzed, and evaluated. The BEE personnel can measure these exposures to routine chemicals found every day on the job in the operational environment to determine the duration, concentration, and the overall risk of the exposure. They can design and recommend safeguards (control measures) to create barriers, such as ventilation systems, personal protective equipment (PPE), and chemical substitutions, to separate toxic substances from the HWS. In addition, they can perform fit testing on RP masks and test the adequacy of ventilation systems in use.

Environmental emissions are another risk to the HWS, as well as to the ecosystem. The BEE can do environmental emissions measurements such as rainwater discharge and air emissions. Depending on the deployed location, this can be a political hot button. Most developed countries have adopted environmental laws that require proper pollution prevention measures to be employed by tenet units. Failure to comply with the host country's demands could jeopardize the mission. The BEE (along with the Civil Engineer Environmental Flight) can reduce the political tension for the forward operating commander by ensuring that local environmental laws are adhered to.

ORM on Physical/Thermal Hazards

The identification of the wide variety of physical hazards requires a diverse and systematic approach using numerous types of measuring instruments. The key to avoid over tasking the limited assets and time spent on this phase of ORM is to prioritize the threats both by impact on the HWS and the probability of occurrence. The following will examine these impacts and likely frequency as well as demonstrate some of the defenses the BEE can recommend.

Noise

"Cumulative overexposure to hazardous sounds and noises cause millions of people to lose their hearing."[9] The constant jet noise on the flightline makes noise-induced hearing loss a likely non-debilitating physical illness to be experienced in the combat theater. "Since no treatment is available to mend noise-induced hearing loss, preventive measures are paramount."[10] Noise dosimetery is an effective way to survey and identify high-risk areas. The BEE can determine the intensity and duration of routine exposures and recommend the level of PPE required for protection or recommend administrative controls such as relocation of certain high-risk shops. On the other hand, an engineering solution may be recommended such as building earthen berms or adding vestibules to augment the sound.

Light

The occupants who experience symptoms such as headaches and eyestrain usually identify poor lighting conditions. The BEE can do lighting surveys on these workspaces to quantify and analyze the cause of the conditions. Typical prescriptions often involve low cost changes such as lowering the lights or moving the illumination directly over the work surface, rather than the more knee-jerk response of just adding more lights. The BEE works closely with CE personnel in the later ORM steps of control decisions, implementation and review.

Lasers and other high intensity lighting devices should be surveyed and inventoried. Operators must be educated in the proper use and precautions necessary to prevent injury. The BEE staff can properly catalog and inventory these useful, but potentially dangerous devices. They can also provide training (control measure) to ensure those individuals operating these devices are aware of dangers from misuse.

Non-ionizing and Ionizing Radiation

Non-ionizing radiation emitters like communications transmitters and microwave stations can be properly placed if BEE personnel are on the advance team. For instance, although operators are aware of the need for certain quantity distances away from the dish, they have no way to measure the affects of their signal in low angle transmitters (e.g. tropospheric shots). Should their dish be at a lower elevation than a high traffic road that intersects their broadcast, they could be inadvertently exposing passersby. The BEE staff can determine with their equipment the exact location of the broadcasting beam and the intensity of the signal at the roadway. Control measures may range from do nothing to relocation of traffic or elevating the dish. The commander and the BEE must work closely to find an acceptable solution that preserves both the mission and the HWS.

Radioactive materials (produce ionizing radiation) need to be identified and inventoried. Strict controls must be implemented on all devices containing such materials. The BEE should be appointed as the Radioactive Materials Officer, because of the qualification required by the position and their unique training and certification in this area. They not only are useful for all parts of ORM in ionizing radiation matters, but can also serve on the Incident Command Team to survey accident sites and determine if radiation leakage has occurred and to what extent. They can provide education to operators of ionized radiation equipment operators, emergency personnel, and to anyone who may unwittingly come in contact with this serious and invisible threat.

Heat Stress Injuries

The probability of heat stress injuries to the HWS increases due to climate, workload, and attire. BEE personnel have the equipment to identify high threat conditions based on relative humidity, dry and wet bulb temperatures, and medically accepted heat stress charts and graphs. They can analyze the conditions and make scientific recommendation as to the MOPP level and work-rest cycles needed to avoid injury to the HWS. Because the entire unit is at risk to heat stress in balmy, hot environments (especially in MOPP 4), constant monitoring of changing environmental conditions is a critical function to maintain operations.

Ergonomic Hazards

Ergonomic Hazards such as RMI and low back injuries must be identified, if possible, before they occur. Fortunately, the general health of the HWS in the armed services reduces the chance of some injuries, however the urgency of some tasks put some individuals at risk. Bomb loaders, Supply and CE personnel, and generous airmen lending a hand to load or off-load aircraft are certainly at risk. The BEE can help to identify and analyze the relative risk for ergonomic injury. They can make recommendations for controls such as the use of dollies, forklifts, or something as easy as using two people instead of one to off-load heavy equipment. The last steps of ORM, implementation and supervision, for ergonomic hazards should be left to the section supervisors.

Minimizing the Risks to the HWS

The BEE can use occupational health controls and surveillance and industrial hygiene to minimize the risks to the HWS. Using this methodology in conjunction with in the Operational Risk Management (ORM) process, critical subsystems of the HWS can be defended. However, with all of the capabilities of the BEE shop, historically they often were not used to their fullest extent or sometimes not at all. The following historical sketch will show some of the preventable illnesses encountered in combat with emphasis on the United States' most recent large-scale contingency, the Gulf War.

Notes

[1] Office of the Under Secretary of Defense, 30 NOV 1995, Memorandum for the Director of the Joint Staff, Subject: Safety and Occupational Health Risks Associated with Deployments to Bosnia.

[2] Ibid.

[3] OSHA 3143, 1994

[4] Walt Schaefer, *Stress Management for Wellness*

[5] AFI 91-213

[6] AFI 91-213

[7] Ibid, 89-94.

[8] Op cit., Bioenvironmental, sec 2-12, p. 53.

[9] Op cit., McCunney, p.230.

[10] Ibid, p.230.

Part 3

HISTORY OF ILLNESS IN COMBAT

Those who do not study history are doomed to repeat it.

—John F. Kennedy

General Effects of Indirect Attacks on U.S Forces in History

From as early as the Civil War, the effect of indirect attacks, such as disease, on U.S. forces has been painfully evident. Diarrheal diseases, primarily dysentery, actually killed more soldiers than combat wounds in that conflict.[1] The lack of knowledge about proper personal hygiene and water treatment were likely culprits to the spread of incapacitating illnesses that dramatically reduced combat potential in earlier wars. However, many other agents are responsible for the degradation of U.S. forces in foreign areas.

For instance, in the Pacific Theater during World War II, Malaria plagued American warfighters. Commanders lost five times as many soldiers to this mosquito borne illness as they did to enemy fire.[2] Dengue fever, another mosquito borne illness, reduced the 1st Marine Corps to 85% in the first month on station in Somalia. There was limited medical intelligence available prior to deployment to this area. In spite of some prophylactic efforts, the troops suffered relatively high casualties. [3] Even with modern immunizations the threat of losing personnel to non-combat related illnesses (NCRI) is still a factor that must be considered and addressed. Examination of the most recent large-scale military operation, the Gulf War, will make a case study demonstrating the need for BEE expertise to assist the commander in performing ORM in the combat theater.

Gulf War Hazards

The Gulf War was characterized by the meticulous buildup of the world's most modern military force in a hot, barren environment. The U.S. forces were far superior to the opposition, however, friendly forces were degraded by their own lack of adequate defenses. Primary non-combat related injuries (NCRI) were exposure to foodborne and waterborne disease, respiratory hazards from painting and oil well fires, and radiation exposures from depleted uranium sources. Many veterans of the Gulf War have reported experiencing a variety of physical symptoms, collectively know as Gulf War illnesses. In response to concerns, the Department of Defense (DoD) established a task force called The Office of the Special Assistant for Gulf War Illnesses in June 1995 to investigate all possible causes of these reports by veterans.[4] The effects of these hazards caused operational degradation and still plague the Office of the Secretary of Defense today.

Biological Illness

Even though extensive prophylactic measures were taken by airmen deployed to the Persian Gulf, many pilots were DNIFed (Do Not Fly status) due to waterborne disease and foodborne illness. Although deployed troops were briefed on public health measures, over 55 percent of the deployed forces were struck with diarrheal disease in the Gulf War during their first month in theater.[5] Of these 55 percent, one fifth of them caused the loss of one or more duty days. It was determined that improper personal hygiene along with unapproved local food consumption likely caused many of the illnesses.[6] A similar degradation of the Human Weapons System was experienced in the American Civil War, in which more soldiers died from diarrheal disease than combat injuries.

Chemical Illness

In the rush to mobilized troops and equipment for Operation Desert Shield and Desert Storm many standard procedures were modified, including the necessary painting of equipment and vehicles from the woodland camouflage of greens to the desert colors of tans and browns.[7] Because of this, almost 500 soldiers were exposed to hexamethylene diisocyanate (HDI), a principle health threat in CARC paint, in spray painting operations in the Saudi Arabian ports of Ad Dammam and Al Jubayl. The exposures were caused by failing to adequately supply respiratory protection and from tasking troops who were not trained to carry out painting operations.[8] "Despite repeated health and safety inspections over a seven-month period (December 1990 – June1991) that identified serious deficiencies and hazards, painting activities continued at these two facilities [Ad Dammam and Al Jubayl] with limited improvements."[9] "Inhaling high concentrations of some of the compounds and solvent in CARC paint can cause some short-term symptoms like coughing, shortness of breath and watery eyes. Long term exposures can lead to respiratory problems, including asthma."[10] Though OSD investigations could not definitively link CARC painting operations to the mysterious Gulf War Syndrome, a number of soldiers that were involved with painting operations have reported adverse respiratory effects from exposures caused by HDI and solvents in painting.[11]

Another unforeseen respiratory hazard experienced by Gulf War veterans was smoke from the oil well fires set by retreating Iraqi troops. "In general, U.S. troops were not well prepared to protect themselves against the acute health effects presented by the oil fire smoke."[12] Though the U.S. troops all had M-17 A1 gas masks, these masks were designed for chemical agents and not the heavy particulate matter caused from oil fire smoke. The oil fire smoke would quickly clog the fine material of the chemical filters. The filters then could not be used for the threat in which they were intended, specifically chemical and biological agents.[13] This forced soldiers to use makeshift protection such as handkerchiefs, cravats, workshop dust masks, or whatever could be obtained through the initiative of the troops.[14] Exact numbers of exposed troops are not available, however with approximately 750 damage or destroyed oil wells; a significant portion of the deployed theater was affected. The concentrations of harmful vapors and smoke were not sampled until May 1991, so it is difficult to calculate the risk level or the effects on U.S. troops.[15]

Physical/Thermal Illness

Radiation, in the form of depleted uranium (DU), was a newly recognized physical hazard that disabled U.S. forces in the Gulf War. Firefighters, security police and rescue personnel were exposed while trying to recover an A-10 aircraft carrying 30 mm DU rounds that crashed at King Khalid Military City in northern Saudi Arabia.[16] Additionally, aircrews correcting "hangfires" from the GAU-8 cannon aboard A-10s with 30 mm DU round were potentially exposed to harmful radiation. Over 30 cases of DU radiation exposure were confirmed and the office of the Special Assistant for Gulf War Illness is still investigating incidents.

"DU's ability to self-sharpen as it penetrates armor is the primary reason why DU is a more potent weapon than alternate tungsten munitions, which tend to mushroom upon impact. Fragments and uranium oxides are generated when DU rounds strike an armored target. The size of the particles varies greatly; larger fragments can be easily observed, while very fine particles are smaller than dust and can be inhaled and taken into the lungs. Whether large enough to see, or too small to be observed, DU particles and oxides contained in the body are all subject to various degrees of solubilization—they dissolve in bodily fluids, which act as a solvent."[17]

Once dissolved, uranium may react with molecules in the body and may exert toxic effects. These effects range from cellular necrosis (death of cells) in the kidney to by kidney. Once dissolved in the blood, the kidney in urine will excrete most of the uranium present within 24-48 hours.[18] The body retains the remaining portion indefinitely. "Insoluble uranium oxides, if inhaled, can remain in the lungs for years, where they are slowly taken into the blood and then excreted in urine."[19]

Recent Employment of Preventive Health Measures

Many illnesses caused by indirect attacks have been experienced in combat. However, with the integration of BEEs and other health technicians, great strides have been made in preventing illness in the combat theater. A detailed analysis of specific problems as well as successes will help to illustrate how to focus future efforts to safeguard the HWS.

Notes

[1] Charles M. Levy, and Truman W. Sharpe, "Medical Challenges for Deploying Forces", Marine Corps Gazette, Feb 97, p.55-56.

[2] Paul L. Andrews, "Malaria: A Threat to U.S. Troops", U.S. Army Center for Lessons Learned, May – Jun 97, http://call.army.mil/call/nftf/mayjun97/malaria.htm

[3] Levy and Sharpe, op cit, p.55. taken from - Mark E. Butler, Force Protection in MOOTW: A Medical Perspective, ACSC Research Report, April 1998

[4] Lessons Learned –Gulf War, http://www.gulflink.osd.mil

[5] Ibid, 55.

[6] Ibid, 58.

[7] Ibid.

[8] Lessons Learned –Gulf War, http://www.gulflink.osd.mil/carc_paint/carc_paint_s08.htm

[9] Lessons Learned –Gulf War, http://www.gulflink.osd.mil/carc_paint/carc_paint_s08.htm

[10] Office of the Secretary of Defense, News Release, No. 089-00, 24 Feb 2000.

[11] Ibid.

[12] Lessons Learned –Gulf War, http://www.gulflink.osd.mil

[13] Ibid.

[14] Ibid.

[15] Ibid.

[16] Ibid.

[17] Ibid.

Notes

[18] Lessons Learned –Gulf War, http://www.gulflink.osd.mil/du/du_en.htm
[19] Ibid.

Part 4

Focusing Preventive Medicine Efforts

Military operational planning should incorporate measures to meet occupational safety and health standards even under constrained conditions.

—Office of the Secretary of Defense[1]

Coordinated Efforts

In the past, lack of a coordinated effort between the commander and Preventive Medicine (PM) professionals to educate enforce PM controls such as proper sanitation and hygiene has led to needless degradation of the HWS and reduction of the available fighting force. The lack of cooperation between the services have also caused some confusion, and in some cases damage to the HWS. The following will analyze some of the problems as well as some of the successes, in the evolution of the incorporation of PM professionals in the combat theater.

Failures in Focusing PM Efforts

The military, in general, could have benefited if more PM professionals, specifically BEE personnel, were sent into theater both prior to and during the Gulf War conflict. This increase in personnel would have allowed commanders to focus their efforts by utilizing these preventive health experts to offset some of the damage to the HWS in the Gulf War. However, there are limitations to how effective PM efforts can be if commanders are not willing to support inspections and recommendations. It is understandable that risk is a part of war fighting, but it is prudent to trust and support the individuals charged with running the PM program, namely the Aerospace Medical Council (AMC), which consists of the Public Health Officer, Flight Medicine and the BEE. Two specific examples of problems in enforcement of PM controls caused unnecessary degradation of HWSs from biological and chemical hazards during the Gulf War.

Biological – Improper sanitation and hygiene

Hand washing and water testing are continual problems in overseas deployments. The enforcement of hand washing, no matter how elementary it may seem, is an important and easy way to avoid serious damage to the HWS. Though the issue of water testing is slightly more complicated, proper prior coordination with BEE personnel can not only assist in eliminating waterborne disease, but can also reduce food borne illness due to contamination from water in the food's preparation. In the Gulf Conflict, improper hygiene along with consumption of food from unapproved local eateries was blamed for much of the diarrheal diseases that struck and incapacitated the HWSs.[2]

Chemical – Inhalation hazard exposures

Because long term side effects may manifest in sensitive individuals, as with CARC paint exposure in the Gulf War, it is imperative for a commander to weigh the risk of his HWS's health and the urgency for accomplishing the mission at hand.[3] Several options are usually available for accomplishing the mission, while still reducing risk to the valuable HWS. Proper safety equipment and training was not provided to individuals exposed to the CARC paint in the Gulf War incident. The activities were identified as having serious deficiencies on health and safety inspections over a seven-month period, yet the commanders of the two facilities failed to correct them.[4] Consequently, a large number of soldiers now are reporting respiratory disorders from the likely exposures.[5] The addition of a local PM specialist at each of these units could have assisted in the selection of PPE, in the engineering of ventilation equipment, and in the institution of education and training programs to reduce or eliminate this type exposure. Of course enforcement is not the only difficulty in focusing PM, the lack of education of the hazards on modern battlefield has caused its share of casualties.

Physical/Thermal – DU exposures

The DU exposures were a result of lack of education and the rush for expedience. Most of the preventable exposures in the Gulf War were caused during aircraft accident responses and aircraft landing with "hung" munitions.[6] These exposures were caused more by a failure to plan for such events and lack of proper response equipment, than a flagrant disregard for safety. These modern physical threats of DU, lasers, and non-ionizing radiation could damage or destroy HWS in future conflicts unless commanders take proactive measures in coordination with BEE personnel to identify and educate base personnel.

Successes in Focusing PM Efforts

Though much of the popular attention in the PM arena seems to be directed at failures in the system, there have been many recent successes. The trend of incorporating preventive medicine experts in theaters, especially in South West Asia and the Balkans, has led to a definite reduction in Non-Combat Related Illness (NCRI) on the HWS.

Biological – Pest management

The Navy Forward Laboratory (NFL) established at the "Marine Corps Hospital" in Al Jubayl early in Desert Shield predicted the possibility of insect-transmitted disease. Accordingly, military entomologists (experts in insect and pest control) sprayed high threat areas months prior to troop occupation. Consequently, researchers only found 32 cases of insect borne illnesses out of a population of 750,000 U.S., British, and Canadian Gulf War veterans.[7] These statistics were far better than the results of insect borne illness in U.S. and British troops deployed to the same region during World War II.[8]

In fact, laboratory analysis registered no cases of typhoid fever, cholera or amebic dysentery.[9] The precautionary steps by PM professionals in the theater were successful in defending the HWS and allowed maximum potential of those systems to be exercised. Another remarkable success was in the area of preventing heat injuries.

Physical – Heat stress

In the scalding, dry dessert of the Middle East, dehydration and heat stress were major concerns. Through the diligent efforts of engineers and BEE staff the water supply to prevent heat stress was adequate. Though there were numerous problems in procurement, treatment and storage, the fact that hydration was recommended by PM professionals and stressed by the commanders, less than 0.3% of the force per week (3 cases per 1000 per week) required treatment at an aid station for heat injury.[10] Strong command emphasis on providing abundant water and adequate acclimatization defended the HWS against one of the most probable health threats in that theater.

Balkans – PM Focus at its best

The effects of having the proper number of BEE and preventive health technicians are apparent in the recent military actions in the Balkans. Since the Gulf War, the AMC was deployed by AFMC in support of the Kosovo conflict. USAFE vice commander, Lt. Gen. Bigert, noted in an address that "not a single mission was lost due to illness throughout the Kosovo conflict. On average, in the past wars to include Dessert Storm, losses were at a ratio of 10 NCRI to each combat casualty."[11] Although it would be impractical to attribute this success strictly to the addition of the BEEs in theater, it is clear that the addition of these PM professionals to the fight will contribute to a reversal of NCRI. Actions in the Balkans point to a correlation of maintaining Preventive Health professionals, namely BEEs, and the general overall health of soldiers on the battlefield. To Focus PM efforts to ensure "Full Spectrum Dominance", incorporation of the BEE career field in the combat theater is essential.

Need for Doctrine Regarding Preventive Medicine and the BEE

Other than the fact that Doctrine compels a commander to operate a certain way, it also enables the commander to have some clear guidance on how to plan. DOD Instruction 6055.1 outlines the roles and responsibilities of the commander to establish a Safety and Occupational Health program. The War Mobility Plan (WMP) for the Air Force, as well as Tables of Organization for other services, has Preventive Medicine specialists specified at multiple organizational levels.[12] However, there is not a cohesive Joint Doctrine for Preventive Medicine (PM). Limited mention of PM practices are in JP 3-11 Joint Doctrine for Nuclear, Biological, and Chemical (NBC) Defense and in JP 4-02 Doctrine for Health Service Support in Joint Operations.[13] Both PM sections in these publications agree that there is a need to investigate water quality and possible sources for biological contaminants, however they neglect chemical and physical threats altogether. They also fail to give any consideration to problems associated with joint operations such as the water treatment in the Gulf Conflict.

The lack of a Joint Medical/PM doctrine caused problems during the Gulf War in the area of water treatment. At that time, there was no tri-service standard for chlorine treatment of water.[14] Some services required 2 parts per million (ppm) while some required 5-ppm residual chlorine.

In trying to maintain the maximum standard of five ppm, the water was over-treated.[15] High concentrations of chlorine may have caused some veterans to experience stomach cramping and diarrhea.[16] The result was that the army was tasked with standardizing the U.S. military water policy.[17] While standardization is probably a good thing, a more holistic approach to U.S. military preventive medicine in the form of a Joint Medical/PM Doctrine would help to avoid problem of standardization in other areas in the future.

Need for a supported PM Program

From the past experiences of illness in combat theaters, it becomes obvious that an active PM program is insufficient without close support from the commander. The following will discuss specific recommendations to assist commanders in conducting ORM and protecting the HWS.

Notes

[1] Lessons Learned, Op cit.

[2] Charles M. Levy, and Truman W. Sharpe, "Medical Challenges for Deploying Forces", Marine Corps Gazette, Feb 97, p.55-58.

[3] Lessons Learned –Gulf War, http://www.gulflink.osd.mil/carc_paint/carc_paint_s08.htm

[4] Ibid.

[5] Ibid.

[6] Lessons Learned – Gulf War, http://www.gulflink.osd.mil

[7] Bernard Rostker, Information Paper – Medical Surveillance during Operation Desert Shield/Desert Storm, DOD, 6 November 1997.

[8] Ibid.

[9] Ibid.

[10] Ibid.

[11] Lt. Gen. Bigert, AFMC Leading Edge, November 1999, p. 5.

[12] Rostker, op cit.

[13] Joint Doctrine Encyclopedia, 16 July 1997, www.dtic.mil/doctrine/jrm

[14] Op. Cit., Lessons Learned.

[15] Ibid.

[16] Ibid.

[17] Ibid.

Part 5

Recommendations

Good doctors are of no use without good discipline. More than half the battle of disease is fought, not by doctors, but by the regimental officers...

—Field Marshall William Slim

Advocating an active PM program

For combatant commanders to protect their forces, namely the HWS, they must commit to the ORM process. By integrating the BEE staff in every phase of deployment from advanced teams to operations and conflict termination, he will be able to maximize the HWS available.

Defense to Biological Threats

The commander should take at least one qualified BEE per wing on any pre-deployments OCONUS. When deploying to a third-world country or an undeveloped area, this individual can begin the ORM process of identifying hazards and analyzing things such as the water supply, sanitary conditions and local restaurant's food preparation procedures. Additionally, he can investigate the hazards associated with the local flora, insects and animals. An adequately manned and supplied BEE shop should be sent on all wing size deployments to maintain constant surveillance of biological threats. Most importantly, recommendations from the BEE should be carefully considered before overruling them no matter how remedial they may seem. It should noted that diarrheal diseases, namely dysentery, are the leading cause of death in the world and future strains may be resistant to current antibiotics. The fact that deployed forces will be exposed to these diseases should warrant serious consideration from the deployed commander.[1] Command enforcement of simple things like proper sanitation and hygiene pays big dividends in preserving the HWS. Commanders not only must enforce hygienic measures, but also empower the AMC to educate base populace on proper sanitary and occupational health measures.

Defense to Chemical Threats

Commanders need to give BEE personnel access to as much of the base as possible in order to inventory chemical storage and ensure proper PPE is available and worn. The BEE also may monitor environmental compliance with disposal of hazardous waste, which will not only please the host country, but will assure that U.S. HWSs do not fall victim to inadvertent toxic ingestion. He and his staff should also have adequate access to computers with Internet linkage in order to locate specific information (such as Material Safety Data Sheets) on substances used in theater to assist in mitigating a spill or accidental release. The commander should ask for a weekly update of the status of preventive and occupational health threats (perhaps at commander's call) so that he and his staff can stay appraised of the health of the HWS.

Defense to Physical/Thermal Threats

The commander and the BEE should work together on the ORM process for identifying and reducing the exposure to physical and thermal threats. The commander can look to the BEE to survey the quantity distances of non-ionizing radiation equipment (e.g. microwave transmitters) as well as locate and inventory ionizing (radioactive) sources. The commander should also rely on the BEE for accurate heat stress information and recommendations for work rest cycles and other mitigating recommendations.

Command Support

Commanders should be held accountable for the health and safety of their forces. They must take appropriate action providing the necessary training, obtaining appropriate safety equipment, and taking immediate steps to resolve any identified safety and health deficiencies. By allowing the BEE staff to analyze and recommend PM measures and options, the commander can decided on selected courses of action and provide the command emphasis to see that the PM measures are carried out. The commander may delegate review and supervision to the BEE, Safety, and individual shop supervisors and have them report findings on inspections and progress on deficiencies.

A Joint Medical/PM Doctrine should be developed in cooperation with all the services' inputs. The development of a holistic approach to U.S. military preventive medicine in the form of Joint Doctrine will help to avoid problems of standardization and provide guidance in the future.

Notes

[1] Ibid, 55-56.

Part 6

Conclusions

"Since then [Dessert Storm], team Aerospace [BEE, PHO, and Flight Medicine] has played a large part in reversing non-combat related illness rates. Good health is a force multiplier, keeping our soldiers in the fight where they are needed."

—Lt. Gen. Bigert, Vice Commander USAFE
During Balkans Conflict

Preventive Defense of the HWS

Much like Trenchard insisted that airpower was inherently offensive; the idea of protecting the HWS is inherently preemptive.[1] Waiting for disease and illness to strike and then treating the symptoms is akin to allowing enemy aircraft to take off before engaging them. In both cases, the cure is much more costly than proactive measures. The HWS can be defended provided there is and aggressive PM program that identifies biological, chemical, and physical/thermal hazards.

In order to fight effectively, all weapons systems must operate at their maximum potential. ORM is a sound process for identifying and evaluating threats and creating control methods to protect these systems. The Human Weapons System is the military's most valuable weapons system and it must be safeguarded. By applying the PM skills of the BEE in the ORM process, deployed commanders will preserve their fighting forces. This career field can identify, analyze, and quantify threats to the HWS as well as provide preventive medicine in the form of surveys, selection of personal protective equipment, water testing, and compliance monitoring. These capabilities translate to maximum operational effectiveness with minimum health risk to the HWS. Using the BEE career field, the commander can shape the battlefield to provide the most ideal conditions for force employment.

Notes

[1] Col. Phillip S. Melinger, <u>The Paths of Heaven</u>, Air University Press, 1997, p. 51.

Appendix A

This section is provided to give the symptoms and duration of illnesses that may be encountered in a combat theater. These are the top nine biological threats as prioritized by the Department of Defense. It is not all-inclusive but may serve as a guide to better understand the operational effects on the war-fighter.

Disease	Incubation Time (days)	Fatalities (percent)
Anthrax	1 to 5	80
Plague	1 to 5	90
Tularemia	10 to 14	5 to 20
Cholera	2 to 5	25 to 50
Venezuelan Equine Encephalitis	2 to 5	<1
Q Fever	12 to 21	<1
Botulism	3	30
Staphylococcal Enterotoxemia (Food Poisoning)	1 to 6	<1

Table 2

Anthrax – also known as woolsorter's disease comes in two forms: Cutaneous (or skin) and Weaponized. The cutaneous anthrax is endemic to some regions and is transmitted to and from the skin or fur of animals. Lesions and ulceration of the effected external areas characterize it. It is usually not fatal. Weaponized Anthrax is a more concentrated form of the pathogen and can be contracted by breathing or ingesting the contaminant. It can be nearly instantaneously fatal to anyone not inoculated against the disease.[1]

Plague – A disease that has a history of reaching epidemic proportions. Transmitted by fleas on rodents such as rats. May be delivered by enemy aerial sprayers or other vector devices. It is characterized by tender and swollen lymph nodes (or buboes) that may burst and usually accompanied by a fever. Treatment is possible if diagnosed early. Nearly all forms of the disease are fatal if left untreated.[2]

Tularemia – Also called rabbit fever or deerfly fever is transmitted by arthropods such as biting flies or ticks, drinking contaminated water, or handling infected animals. Its symptoms are swollen lymph nodes and usually primary ulcers on the bite site. Clinically, it may be confused with plague.[3]

Cholera – An acute bacterial disease with sudden onset of profuse painless watery stools and rapid dehydration. Is usually contracted by drinking water contaminated with feces or vomitus or ingestion of foods prepared with dirty water, dirty hands or that has had contaminated flies on it. In severe untreated cases, death may occur within hours.[4]

Venezuelan Equine Encephalitis – Mosquito borne illness with symptoms that resemble influenza with chills, severe headache, fever, nausea and vomiting. Symptoms usually last 5 days.[5]

Q Fever – A disease delivered by dust from infected feral rodents and farm animals. It is characterized by onset of sudden chills, weakness, malaise, and severe sweats. The disease is endemic to all continents.[6]

Botulism – A highly lethal substance that can occur naturally in contaminated food or can be weaponized. Minute amounts of weaponized botulism cause rapid paralysis and death. Iraq possessed stores of this toxin during the Gulf War.[7]

Staphylococcal Enterotoxemia (Food Poisoning) – An intoxication (not an infection) of abrupt and violent onset of severe nausea, cramps, vomiting, lowered body temperature and lowered blood pressure. Toxin is transmitted in foods such as poorly cured ham, pastries, salad dressings, and meat products that have come in contact with food handler's hands without subsequent cooking.[8]

Other diseases affecting deployed HWS:
Dengue Fever – disease transmitted by mosquitoes that has a 3 to 14 day incubation period. It is characterized by the sudden onset of fever that lasts from 5 to 7 days with skin rash, intense headache, anorexia and gastric-intestinal disturbance. Epidemics are explosive (400,000 affected in Cuba, 1981) but fatalities are rare without hemorrhagic fever present.[9]

Diarrheal diseases – Any number of illnesses that produce symptoms of frequent loose or watery stools, and often are accompanied by vomiting and fever. Diarrhea is a symptom of infection by bacteria, viral and parasitic agents. Seventy to Eighty percent of reported episodes are from people visiting treatment facilities in less-developed countries. Dysentery, a common diarrheal disease, is signified by scanty stools containing blood and/or mucus and can be persistent (lasting longer that two weeks).[10]

Notes

[1] Abram s. Benenson, <u>Control of Communicable Diseases in Man</u>, American Public Health Association, 15th Edition, 1990, 17-20
[2] Ibid, 324-327.
[3] Ibid, 466.
[4] Ibid, 89-94.
[5] Ibid, 37-39.
[6] Ibid, 350-352.
[7] Bernard Rostker, Information Paper – Medical Surveillance during Operation Desert Shield/Desert Storm, DOD, 6 November 1997.
[8] Ibid, 170-171.
[9] Benenson, op cit, 117-122.
[10] Ibid, 129-131.

Glossary

ACSC	Air Command and Staff College
AMC	Aerospace Medical Council, which consists of the Public Health Officer, Flight Medicine and the BEE
BEE	Bioenvironmental Engineer
CE	Civil Engineer
DOD	Department of Defense
DU	depleted uranium
HWS	Human Weapons System
NCRI	Non-combat Related Illness
ORM	Operational Risk Management
PM	Preventive Medicine
USAF	United States Air Force

Diarrhea: a potentially epidemic problem in field conditions. Symptom of many communicable illnesses. Causes loose bowel movements and rapid dehydration.

Heat injury: one of the most significant health threats early in the Gulf War deployment. Any illness brought on by heat stress such as heat exhaustion and heat stroke.

Injury/musculoskeletal conditions: a major cause of lost man-days from training and deployment activities.

Laser: Any of several devices that convert incident electromagnetic radiation of mixed frequencies to one or more discrete frequencies of highly amplified and coherent visible radiation.

Microwave: Any electromagnetic radiation having a wavelength in the approximate range from one millimeter to one meter, the region between infrared and short-wave radio wavelengths.

Prophylaxis, prophylactic. Preventive measures such as inoculations, immunizations, etc.

Radar: A method of detecting distant objects and determining their position, velocity, or other characteristics by analysis of very high frequency radio waves reflected from their surfaces.

Respiratory conditions: colds, pneumonia and other respiratory problems are common and can be widespread during any deployment.

Eye problems: eye infections, like "pink eye," can be epidemic in field conditions, also corneal abrasion from blowing sand was a risk in the desert.

Unexplained fevers: an unexplained fever may be the first sign of diseases, such as sand fly fever, malaria, and other serious infections.

Psychiatric conditions: the stresses of deployment and combat often cause psychiatric symptoms.

Bibliography

Andrews, Paul L. *"Malaria: A Threat to U.S. Troops"*, U.S. Army Center for Lessons Learned, May – Jun 97, http://call.army.mil/call/nftf/mayjun97/malaria.htm

Benenson, Abram S., Control of Communicable Diseases in Man, American Public Health Association, 15th Edition, 1990.

Bigert, Lt. Gen. AFMC Leading Edge, November 1999, p. 5.

Bioenvironmental Engineering Applications and Principles, USAF Directorate of Aerospace Safety, Norton AFB, CA, 1963.

Breyfogle, Bill, Major, *"Lessons Learned: Operations Other Than War"*, Engineer, vol.24, p. 23-25, April 1994.

Cook, Wayne, *"Radioactive Material and You"*, Army Logistician, vol. 30 no. 3, p. 36-37, May-Jun 1998.

Clayton G. and Clayton F. (Eds.) Patty's Industrial Hygiene and Toxicology: Toxicology, (4th Edition). New York: Wiley, 1991-1994.

Goodman, Sherri W., *"Safety remains a top priority in today's military: Remarks at the Defense Agency Safety and Occupational Health Symposium, Fort Belvoir, VA"*, Defense Issues, vol.13 no. 24, 1998.

Pinker, Steven, How the Mind Works, W.W. Norton and Company, New York, 1997

Koloff, Phillip, Captain, *"Bioenvironmental Engineering Controls – Protecting the Worker and Reducing Exposures"*, Mobility Forum, vol. 8 no. 3, p. 23-24, May-Jun 1999.

Lessons Learned –Gulf War, http://www.gulflink.osd.mil

Levy, Charles M. and Truman W. Sharpe, *"Medical Challenges for Deploying Forces"*, Marine Corps Gazette, Feb 97, p.55-58.

McCunney, Robert J., A Practical Approach to Occupational and Environmental Health, 2nd Edition, Little, Brown and Company, Boston, MA. 1994.

Meilinger, Col. Phillip S. The Paths of Heaven, Air University Press, 1997, p. 51.

Office of the Under Secretary of Defense, 30 NOV 1995, Memorandum for the Director of the Joint Staff, Subject: Safety and Occupational Health Risks Associated with Deployments to Bosnia.

OSHA 3143, 1994

Rostker, Bernard, Information Paper – *"Medical Surveillance During Operation Desert Shield/Desert Storm"*, DOD, 6 November 1997.

Schaefer, Walt, *Stress Management for Wellness,* ACSC Coursebook Vol. 1, 1999.

Skier, Major, *"How the BEE can assist in an OSHA inspection"*, TIG Brief, vol. 43, no. 5, p. 10, Sep-Oct 1991.